I dedicate this book to all the scientists of humanities past who have collectively labored in an effort to bring comfort and control to all our lives. I believe I have followed in their footsteps and to the best of my abilities lived a life for the betterment of mankind. I thank my entire family for being the rock in my life. I thank my son Austin for just being who he is- my son.- For without question if I have not understood the miracle of life through our children- I would not have continued to labor to such a degree of sacrifice for humanity under the realm of the Heavens. I Thank God for all things.

The year was 2002, I was still in the midst of a gravity experiment. The picture above was addressing gyroscopic action- I still feel gravity manipulation is by the little toy top (Inertia) –and as I was reading my favorite science book- The sourcebook on atomic energy by Samuel Glasstone- and the construction details of pressurized water reactors- the idea simply hit me like a lightning bolt- interesting- as the lightning bolt has always been my trademark as shown on my shirts as I was a self employed master electrician.

Lightning Fusion

"The Marriage of Hot and Cold Fusion"

By: Solomon Sami Azar

www.noblefuse.org

2nd edition

I offer a completely new approach in achieving an abundant energy output via the nuclear fusion process. This paper is an elaboration of my patent registered with the United States Government titled "Nuclear Fusion Generator"US60/840,493 or NFG system. In the title of this paper where I pronounce the statement of "The Marriage of Hot and Cold Fusion", I desire to offer the reader in what mode of science does this new approach takes in which I have originally named in 2002 "Electro-Magnetic Fusion" to I believe more appropriate "Lightning Fusion". In the following pages it shall be apparent that indeed what is being presented has the attributes of both classes of nuclear fusion devices in that; in similarities to hot fusion, it is this authors fundamental belief that in striving for magnetic pinch qualities of relativistic charged particles that in itself is the cause of the actual nuclear combination of elements as opposed to simply being poloidal currents of magnetic strength in order to be contained by external magnification within vacuum vessel with sole intent of increasing temperature to that comparable to our Sun; the similarities to Cold fusion are merely that it is this authors use of the liquid state of water is used for several reasons which will be stated later, however, in which will be noted now that this mode of application has nothing to do with established theories of Cold Fusion to date, in fact it is this authors opinion that what is new and novel being presented will actually describe the superfluous positive results of Cold Fusion and in fact are the result of the modes of excitation I believe are generated when electrical current traverses a liquid hydrogenous solution and not at all with the belief in the "catalytic" cleaving actions of various electrode materials used in Cold Fusion cells.

Thus, I propose an entirely new ideology of addressing the feasibility of commercial nuclear fusion. My focus is upon the forth state of matter of plasma within the fuel itself water, thereby deriving the name Lightning Fusion. Water itself may be examined in its common forms of solid liquid and gas, it is my intention to offer the reader the next state of matter being an ionized state of plasma within a controlled environment of a man made vessel containing the fuel water.

I have performed an experiment never done before in science- I used a Tesla coil for its use in high voltage high frequency and apply its discharge plasma not upon the dielectric of free air- but to the dielectric of water itself- specifically I used ultrapure reagent grade water from manufacture NERL-this is to establish the high degree of insulation needed for plasma (you cannot have contaminants for conductivity)- I doped my water with heavy water from the manufacturer UNITED NUCLEAR- I built my 1 million volt Tesla coil entire tunable- every aspect of it- as it must be done to TUNE THE OUTPUT DISCHARGE OF THE TESLA COIL to the water itself- once the arc is stable- the voltage may be increased- I have that a prerequisite of 750 kv is needed as an electric field gradient about the charged particles used in fusion(in this case the hydrogen bound in the water molecule) because of voltage drops as expected as in all electrical systems upon the load (load here is the water)- a much higher voltage is needed in order to distribute the voltage gradient upon entire arc plasma length between electrodes in water- THUS- THE HIGH THE VOLTAGE- THE BETTER- within my website you will find a link to a video showing my primitive experiment- BUT MAKE NO MISTAKE ABOUT THIS- THIS IS THE FIRST TIME EVER DONE BEFORE - I propose nuclear fusion of water/heavy water- my little experiment IS THE ROAD TO NUCLEAR FUSION- we must universally connect the dots- put two and two together- and conclude this- MY EXPERIMENT MUST BE REPEATED ON A LARGER SCALE- my system is a direct replacement of nuclear power plants particularly of the pressurized water reactor which uses heavy water already - a vessel already built for gamma radiation and other high energy flux which will emit with the plasma arc-

Power reactor in which the heat is dissipated from the core using highly pressurized water (about 160 bar) to achieve a high temperature and avoid boiling within the core. The cooling water transfers its heat to the secondary system in a steam generator. Example: Grohnde Nuclear Power Plant in Germany with an electrical output of 1,430 MW.

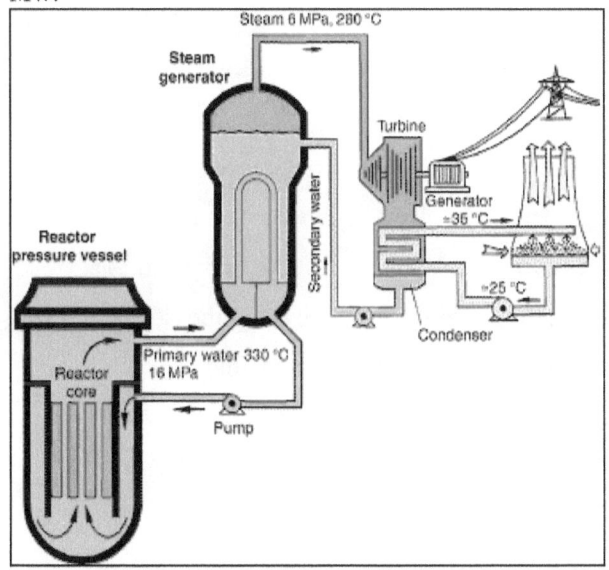

Replace the rod assemblies and use electrodes to conduct the Lightning bolt! High Voltage High frequency will create the magnetic pinch to slam the isotopes of hydrogen together which is bound in water- helium and oxygen are the outgassed products recaptured by expansion tanks-it is absolutely perfect!

_ I HAVE FOUND THE TRUE PURPOSE OF THE TESLA COIL- the answer was always in the lightning bolt- understanding of gamma bursts from lightning discharges have only been recently vindicated from satellite in late 90s---

In what is to follow in this paper, the reader is asked to keep this in mind; I have built an exceptionally crude NFG system and ACHIEVED POSITIVE EXCESS ENERGY. Let us first prime ourselves with the general background of Fusion of nuclear material. The nuclear fusion process offers the world a practically unlimited fuel supply that is also clean to the environment with no greenhouse gas emissions. It is known that to attain fusion of nuclear material, the "easiest" of all the available elements would be Hydrogen and its isotopes. Hydrogen with a subscript of H has only a single proton in its nucleus; the vast majority of hydrogen bound in water also has no neutron attached. If a neutron was attached it is called a deuteron or simply "D". If yet one more neutron were attached as the third possible stable isotope of hydrogen (12 yr half life), its name is tritium or simply "T". Hydrogen is seen as the candidate for fusion as opposed to all other elements because it has only one positive charge being the single proton. Hence, in trying to "squeeze" two hydrogen's together the fundamental force needed is derived from Coulombs law;

$$F = k_C \frac{|q_1||q_2|}{r^2}$$ (equation 1)

where:

F is the magnitude of the force exerted,

q_1 is the charge on one body,

q_2 is the charge on the other body,

r is the distance between them,

$k_C = \dfrac{1}{4\pi\epsilon_0} \approx$ 8.988×10^9 N m^2 C^{-2} (also m F^{-1}) is the **electrostatic constant** or **Coulomb force constant**, and

$\epsilon_0 \approx$ 8.854×10^{-12} C^2 N^{-1} m^{-2} (also F m^{-1}) is the permittivity of free space, also called electric constant,

For the fusion of two hydrogen atoms, q1 and q2 is simply one charge value each as there is only one proton each interacting with one another. The amount of energy imparted unto the nuclei of hydrogen to penetrate the coulomb barrier is achieved via temperature. In particular the isotopes of hydrogen both D and T has been found to require less energy to surmount the barrier than hydrogen itself, by using the following understanding of the "gas equation"

$$\frac{1}{2}mv^2 avg = \frac{3}{2}kT$$ (equation 2) (k here is Boltzman constant, T is

temperature)

Kinetic energy depends entirely on temperature. Thus, with the known energy value needed to overcome the coulomb barrier, a temperature value of approximately several billion degrees is required. It is this understanding of the Hot Fusion plasma science that by increasing the temperature of a particle via gas equation 2, that enough energy will be imparted unto the nuclei for it to penetrate the Coulomb barrier. The plasma state is intensely hot and any material in its locale would be vaporized. The plasma state once derived has been contained and forced into the center of vacuum vessels by magnetic fields which hold the plasma current which possesses its own magnetic field derived from an electrical current induced within it as first described by W.H. Bennett in 1934 as a mode of plasma containment dubbed the "pinch effect". Most vessel designs are built upon the concept of the Tokamak (Russian for toroidal chamber in magnetic coils) its shape is that of a toroid with nothing in its center to interact with the plasma channel because of intense heat. Modes of excitation of plasma are entirely derived externally from the toroid via transformers and other forms of additional energy are injected into said plasma such as; lasers, electron beams, and rf injection. It is desired that with such a high magnitude of energy via temperature in the system, a percentage of some collisions occurring between fuel atoms will fuse together and form a more stable atom of helium. This process of forming a more stable atom will create an exothermic energy output via the fusion reaction which is universally accepted. This energy is absorbed by the walls of the vessel. The heat is extracted from vessel walls by coolant fluid to heat exchangers. This heat will produce steam and subsequently electric generation.

There are various types of fusion machines just cited which employ the use of intrinsic magnetic pinching qualities of induced electrical currents, all of which have one fundamental characteristic in common in that the fuel used is in a gas state. This is logical as pure forms of hydrogen or its isotopes are in gas form. Hence, hot fusion machines have been designed around the discharge of energy within a gas state. However some forms of fusion machines use solid fuel pellets of D-T mixtures used in an inertial confinement system in which multiple laser beams are used to compress the fuel into a plasma state for fusion, or highly accelerated charged particles are diverged to a common center focal point as in the Fusor system. All of these noted systems would be in the generally labeled Hot Fusion. On the opposite end of the spectrum from hot fusion is cold fusion. This is akin to the reality of someday achieving superconductivity at room temperature. Cold Fusion uses liquid water and an electrical current is passed within it, it has been deduced that a nuclear fusion process is achievable with the use of electrode material which might posses the ability to act as a catalytic mediator for fusion to occur at a temperature far lower than the energies required in the plasma science fields of Hot fusion.

In my system I shall dispense with the need for temperature requirements altogether and the use of a gas state as a fuel medium. This gas state is a balance between the radiation pressure of the plasma state itself wanting to blow itself apart and balanced by the intrinsic nature of the magnetic field of its induced current ("Bennet pinch") and applied external magnetic fields. .

In my NFG system I fundamentally agree upon the use of fuel in a liquid state for several reasons: A far greater density of fuel available in the liquid state. Though it may be found that various liquids of different elements may be used as the fuel in the NFG system, I shall use water as the starter material because of its intrinsic simplicity of handling, abundance, and safety. Water is a molecule of H20, composed of 2 parts hydrogen and one part oxygen. Various concentrations of heavy water may be used for introduction of deuterons. It is in the liquid state up to a temperature of 374 Celsius and pressures may be attained in modern day vessels as high as 3500 psi such as in coal fired reactors. Present day pressurized water fission reactors use approximately 2600 psi to surround and cool the fuel rods of these fission assemblies. It is desired to have said NFG-system be a turn key device in which any fission power plant built under a pressurized system may simply have its fuel assemblies removed and have presently described NFG system to replace it and began a fusion process, leaving all surrounding hardware of fission apparatus essentially the same such as cooling and heat transfer mechanisms. Water as the fuel has the advantage of not only being the fuel itself, but also the moderator in temperature via the heat exchange system itself and subsequently for steam extraction and electric power generation. In existing magnetic fusion devices, the gaseous fuel of hydrogen which are then energized into the plasma state are controlled by its own magnetic field via an induced current (the Bennet pinch) and external magnetic fields for confinement. In my present system, the same basic principle applies, however it is desired to take the pinch to a far greater level, also, there will be an additional component of control I shall refer to as "liquid compression". Let us begin to envision the device I describe as the "Nuclear Fusion Generator" or NFG.

FIG 1. Vessel housing fuel and fusion assembly

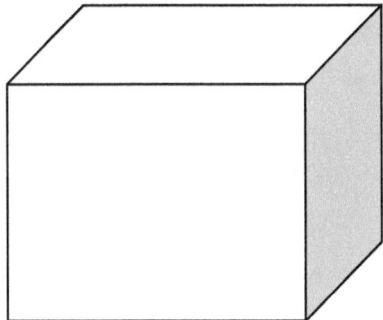

Reference to fig1. Imagine a large vessel of an arbitrary rectangular dimension containing the fuel water. Imagine the vessel to be large enough to also contain in its center an electrode assembly for arc discharge. Imagine this vessel to be large enough that there is sufficient distance from any wall to electrode assembly within center of vessel. Hence, ignore the walls of the vessel for now, however, understand it is filled with liquid water.

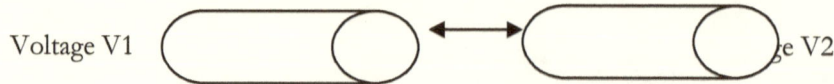

Voltage V1 ... e V2

Fig 2. Electrode assembly

Reference to fig 2 is a simple electrode assembly. Imagine in the center of this vessel two opposing electrodes. Each electrode is appropriately connected to the ends of a power source which supplies the electrostatic potentials V1 and V2 to the connected electrodes. There is a separation of distance between the electrodes facing one another. In between the electrodes facing one another is the dielectric medium of the fuel water which fills the entire vessel and maintained under pressure. Imagine a sufficiently high voltage applied from the voltage source to each electrode via V1 and V2 to initiate a plasma discharge through the dielectric fuel medium being water. Imagine this discharge to be great enough that the flow of plasma current is via a complete breakdown of fuel material water, or more appropriately the molecule of 2 parts hydrogen and one part oxygen. The source of the voltage is alternating current for several reasons. One; alternating currents will eliminate the gaseous amounts of elements occurring at each electrode as in an electrolysis system, for example whereby at an instant in time when one electrode is a cathode, negative relative to the opposing electrode being the anode, hydrogen gas would accumulate. Because of alternating current, the electrodes alternate between being cathode and anode, hence, the appropriate gas will be formed at each cycle, and an opportunity for recombination will result. Secondly and more importantly, because of inductive properties at higher frequencies, a sufficiently high enough frequency is required to "resonant" with the plasma discharge itself. It is known that additional magnetic qualities such as skin effect are reached as higher frequencies are obtained. This is understood as the magnetic energy component of the oscillating field has increased. This shall be elaborated later. Also, in regards to thoughts of possible "outgassing" of free hydrogen or oxygen which might occur and are within the vessel itself, it is realized that during plasma conduction of such a high amplitude of power, a large amount of electromagnetic energy in the form of x-rays and gamma rays will also be emitted, hence, a large activation energy potential resides within the vessel to "recombine" the out-gassed products into its more useable form of a water molecule. Of course, during its recombination back to a water molecule, the exothermic reaction will be absorbed by the medium, and hence, no energy is lost but contained in the vessel.

Reference to fig 3. This is a Tesla coil showing an Rf arc discharge from its discharge terminal and traversing thru air to primary ground on the power supply. This is an example of the type of power source required in order to initiate a plasma discharge between said electrodes and traversing thru the dielectric fuel water. Typical Tesla coils are known to reach potentials in excess of several hundred thousand volts and "relatively" high frequencies usually in the hundreds kHz range.

FIG 3 Tesla coil with rf current shorting to ground

Reference to fig 4: Because of the high voltage between electrodes within the vessel containing the fuel water, dielectric breakdown occurs within water and plasma current has been established between electrodes in which I refer to as being contained within an entire "Plasma Channel". This Plasma Channel will consist of three distinct zones as noted in figure 4. The plasma zone at the very center will consist of free positive ions and negative electrons. Its temperature will be very hot as all plasma's are by there very nature, thus, it is understood that at this center region, only ionized particles are traversing throughout the plasma zone. If we now move away from this center region of the plasma zone, we quickly find a region much cooler in temperature than the conducting portion of the plasma discharge. This zone I refer to as the gaseous zone. Because the temperature is still higher than 374 Celsius next to the plasma discharge, the fuel water will be gaseous in this area immediately adjacent to the discharge. H2o in the gas state will be found along with free hydrogen and oxygen. In continuing our distance away from the center of the plasma channel, the temperature is yet lowered to a point in which the molecule of water is now in a liquefied form(less than 374 Celsius). This zone shall be referred to as the liquid zone, and it continues to the walls of the vessel. Hence, as there have been described three distinct zones within the vessel containing the plasma channel, there are accordingly two distinct boundary layers evident; first is the boundary layer between the plasma zone itself and the gaseous zone of; secondly, there is a boundary layer between the gaseous zone and the liquid zone. These two "boundary walls" are obviously transition points from one state of energy to the next. Its approximate distance from center of plasma channel will be a function of various parameters such as overall pressure applied to liquid fuel and amount of RF discharge current applied to the system. Let it be understood that the entire volume of fuel e.g. water, is held under pressure. Hence, the gaseous zone is also in contact with the liquid zone and its pressure will be the same accordingly; therefore the pressure "pushing" against the actual plasma discharge will also be this same pressure.

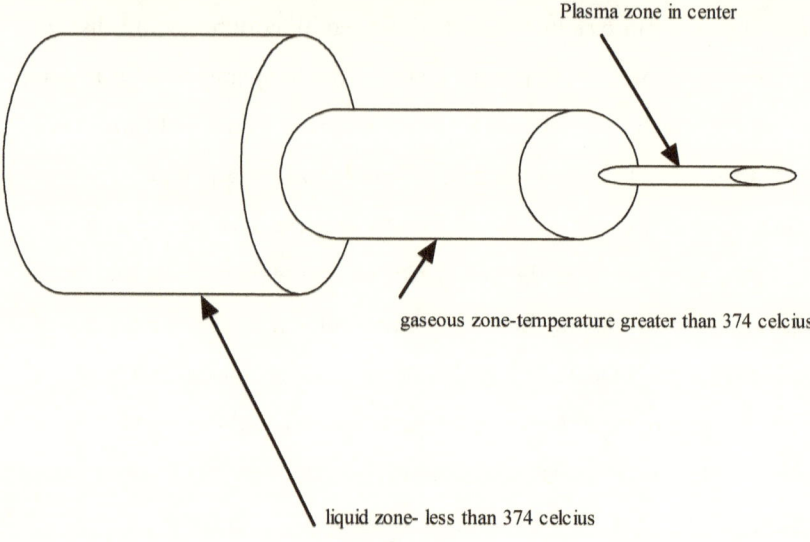

Figure 4. cross section of plasma channel

Plasma zone in center

gaseous zone-temperature greater than 374 celcius

liquid zone- less than 374 celcius

Again, I disregard the temperature requirement for fusion of atomic nuclei which would provide the kinetic energy for the protons of hydrogen to surmount the coulomb barrier. Instead I rely on the magnetic pinching qualities of self-induced currents to overcome the coulomb barrier. My reasoning is as follows:

The velocity of a charged particle in a vacuum with an applied electrostatic field is generally derived from kinetic energy considerations

$$\frac{1}{2}mv^2 = qV \text{ (Equation 3)}$$

whereby m is the mass of the particle (proton or electron), v is the velocity of the particle, q is the value of the charged particle in coulombs, and V is the voltage applied, therefore to derive velocity

$$v = \sqrt{(2qV)/m} \quad \text{(Equation 4)}$$

Hence, the velocity of a charged particle is a function of the voltage V applied to it. For example. If electrons are at a cathode region in which it has been heated to the point of being "boiled off", and a subsequent voltage was applied between cathode and anode, an electron will be accelerated off the cathode toward the anode with a velocity component that is directly related to the equation just cited. This understanding has been used since the early days of electric control in devices such as cathode ray tubes. In regards to an electron, this equation is applicable until the kinetic energy of the rest mass of the electron is reached, 0.511mev, after this amount of energy imparted unto the electron, the relativistic mass equation must be used and accordingly different values of velocity, induced magnetic fields, and mass must be considered.

If now a charged particle has a velocity component v, it now has an intrinsic magnetic field of the value B as a function of velocity

$$B = k' qvX \, \hat{r}/r^2 \text{, (equation 5)}$$ units in S.I. are Tesla for magnetic flux density

the term r^2 is the magnetic field envelope about the charged particle q and the unit vector r-hat is a point in space from point charge q in which the magnetic field may be calculated. One may simply understand that the magnetic density B is directly related to the charge q velocity in space.

where k'=u$_o$/4π or magnetic permeability

Reduced to $B = k'qv$ (equation 6)

it is known that the force on a moving charge in a magnetic field is

$$F = qv \perp B = qvB\sin\alpha \text{ (If not perpendicular) (Equation 7)}$$

It is realized that the magnetic pinch effect is the result of charged particles traveling "side-by-side" next to each other.

It is known that since the magnetic field of each and every charged particle is

$$B = k'qv, \text{ (equation 6)}$$

And the force upon a moving charge q caused an external magnetic field is

$$F=qv \times B, \text{ (equation 7) (x is the cross product)}$$

If the external field B is the field of another charged particle traveling next to another charged particle, then the force between two charged particles via their induced magnetic fields is thus derived as

$$F = k'q1q2\frac{v^2}{r^2} \text{ (Equation 8)}$$

This equation is almost identical to Coulombs law.

Thus, I shall equate equation 8 to Coulombs force equation

$$F = k'q1q2\frac{v^2}{r^2} = kq1q2/r^2 \text{ (Equation 9)}$$

$$v^2 = \frac{k'}{k} \text{ (Equation 10)}$$

In this case since both q1 and q2 which may both represent two electrons adjacent to one another, or two protons, accelerated by the electrostatic potentials set forth by the electrodes stated earlier, then v1 and v2 will be the same velocity so then it becomes v^2(more specifically, v^2 is the vector cross product of the velocities between charges, its magnitude seems to represent the velocity squared). It is with these equations in which I develop an insight into equating the Coulomb barrier force with relativistic magnetic fields of the $v^2 = \dfrac{k'}{k}$ term. I believe with relativistic speeds a magnetic flux density of such magnitude in conjunction with fuel medium at optimum conditions a potential to overcome the Coulomb barrier exists.

The velocity equation no. 5 given forth earlier is entirely correct and subsequent values of velocities are found in a vacuum for charged particles such as in cathode ray tubes. However, since, current in this case is a plasma discharge thru a medium, a vast majority of collisions will occur as in a conductor and there will be an average "drift velocity" in the plasma as in solid conductors, though not the same. The point here is that this velocity equation is the highest attainable velocity for a charged particle if left to travel without hindrance in a vacuum. However, it is realized that "some fraction" of charged particles will attain these velocities within the plasma at some moment in time merely by chance and probability, thus some particles will reach this speed and attain a magnetic field value as a function of velocity. This point must be reiterated further, magnetic pinch devices such as z-pinch or poloidal currents in Tokamaks both use the magnetic pinch ideology, however, it is used to a point to magnetically hold the plasma as other modes of heat injection may occur as the proponents of these technologies fundamentally believe in the temperature requirement for barrier penetration. There is an obvious distinction between what is already employed in fusion devices today and what the NFG system is, and that is the strength of the electrostatic tension I want to introduce into said NFG system. I desire a far greater voltage which will impede a greater velocity upon charged particles; in particular, it is my intention to obtain relativistic speeds of charged particles. Whereby in doing so shall increase the strength of its magnetic field to such a point that fusion reactions may occur. Frequency effects of applied voltage upon velocity must be included and stated later. There is now a force of magnetism between particles which posses a velocity component. It is this very nature of a relativistic velocity-induced magnetic field in which I explore as the "additional" component, amongst others, as a method to overcome the Coulomb barrier. It is my direction to equate the magnetic pinch effect of relativistic charges to that of the Coulomb barrier

Therefore, it seems evident to me that by achieving relativistic speeds, the v^2 term, a magnetic force between charges exist which may cause the nuclei of an adjacent charge to attract one another.

$$v^2 = \frac{k'}{k}$$ It is known that the square root of the electric permittivity k divided by the magnetic permeability k' is equal to the speed of light "c"

It is applying this ideology in a liquid state with the previously mentioned benefits of liquid control and use which makes this invention new as opposed to the prior art of fusion devices of magnetic pinch types or z pinches. Thus, this is not hot fusion in which I rely on temperature as a mode of nuclear combination; I call this "Lightning Fusion", as I seek fusion via a strong electric field which will in turn induce a strong magnetic field for pinching. My road traveled to this magnetic pinch effect was entirely derived in my search for questions regarding gravity. I would like to offer my personal views to demonstrate my line of thought which gives me my reasoning for this fusion attempt in this said method of operation, and in light of such a new approach for fusion with such a high magnitude of global demand, additional information which might seem relevant for this exact fusion device might prove helpful in an understanding of the approach taken. Therefore, the following are excerpts of a paper entitled "On the origin of relativity" in which I wrote in the late 90's. It shows why I believe relativistic velocities can equate the Coulomb force law to magnetism, and therefore offer another method for fusion reactions at the nuclear level.

■■■
"On the origin of relativity"

My views are as follows with this question in mind. Electromagnetic Wave or Corpuscular Photons? It is known that the scientific data gathered via experiments and observation fall with equal weights under a perfect bell curve for a bootstrap wave described by Maxwellian equations or for the classically applied corpuscles "elastically" held to idealized blackbody vibrators. I believe in the latter.

I begin with the special theory of relativity to start the argument.

1. $E = m^* c^2$ Energy of radiation has been equated to mass. The energy of radiation "given off" by the body of mass is equivalent to the mass loss of the body, conversely the energy, E, of radiation "absorbed" by the body is "converted" to mass of the body. These statements are fundamental to the whole subject of atomic physics. Thus, by Einstein, "The mass of a body is a measure of its energy content".

2. $M = Mo/((1-v^2/c^2)^{1/2})$ "The relativistic mass equation". The relativistic mass M is a function of its rest mass, Mo, and its velocity in space.

3. ***The most fundamental equation for magnetism is B = $(q^*k'^*v)/r^2$, whereby $k'=u_o/4\pi$. u_o is the magnetic***

permeability. The magnetic field, B, is a function of an entity q's velocity thru space.

4. M=Mo when v=0: *A very important realism is seen here and yet not fully appreciated. Mo is the rest mass. Though it is called "rest mass", the "rest mass" is not at all at rest. It has an enormous value of energy "intrinsic" to it.* Truly all states are relative. Relative to Earth we have zero velocity when sitting and accordingly some value when we move about. Relative to the surrounding space encompassing the Earth we travel at a large velocity without our slightest knowledge via the Earths spin on its axis of approximately 1,000 miles per hour. Relative to the sun, we travel at approx, 72,000 miles per hour. Our solar system has a spin about the galaxy of approx 600,000 miles/hour, our galaxy has a spin about its own galactic center, and our galactic cluster has a velocity around the central point origin of the universe. The overriding and most precedent of these "relative" velocities is the outward expansion of the universe dubbed the expanding universe and theorized as a result of a big bang explosion hurtling all mass outwards from "center". This overriding and most precedent of velocities I shall name the *Expansion Vector*, for it surely has velocity and direction.

5. Mo=M' $/((1-v^2/c^2)^{1/2})$ Thus, the "rest mass" of now is the result of some primordial mass M' prior to expansion velocity, **for we cannot simply ignore it**. *It must be understood* now that although many "relative velocities" are taken place, only the most precedent and largest of the velocity components is considered. All mass, including earth, is traveling through space at the expansion vector. All smaller velocities are relative and not important in this paper. For example, in the understanding of electricity, if 100 volts were impressed onto a conducting wire and made to perform work, the power is a function of volts times current as depicted by Ohms law. It is inconsequential if on that conducting wire "other voltages" of smaller value (rule of thumb in electrical power is less than ten times can be

essentially ignored such as a 5 volt noise component or a 100 millivolt noise component of different frequency). The analogy being all smaller velocities relative to space are insignificant to the overwhelming velocity of the expansion velocity.

6. $E = h*f$, and $m*c^2 = h*f$, and then $m = h*f /c^2$: **Postulate one**: *The fundamental entity is the corpuscle, or quanta in which I shall refer to hereafter, is when $f=1$, $m=7.37* 10^{-51}$ kg.* Thus, a proton contains $2.27* 10^{23}$ quanta (mass of proton / mass of quanta). An electron, or cloud, or orbital, contains $1.24* 10^{20}$ quanta (mass of electron / mass of quanta) I use $f=1$ for it must be the fundamental quanta, any multiple of f such as $f=10^{14}$ or visible light is implying the atom oscillate at 10^{14} and at each oscillation quanta is broken from its "layers".

7. These "layers" are in fact related to (though not exactly) the circular orbit layers, "n", which are neglected in Bohr's theory as a physical reality but critical for another purpose, and also the radius values in quantum mechanics are closely tied to these "layers", **more appropriately via mass-spring systems**. Bohr noted himself that it was just as applicable to note circular "stationary" orbits of layers were possible.

8. **Postulate Two**: *Each individual quanta has an intrinsic magnetic field about it via the expansion vector.* Magnetism by definition is the velocity of q in space. A fundamental understanding must be appreciated here in that any q with a velocity to space will have a magnetic field with respect to space. This magnetic field is a deformation of space. To describe a magnetic field it must encompass equally its effect on the surrounding space (r^2 about entity q). All intrinsic spin as spoken about in regards to electrons and such is the magnetic field and is inherent via all entities traveling thru space at a very large velocity. This field is a function of some fundamental "primordial

entity" q, with a velocity in space. Space will be distorted about q as a function of Uo, a natural "resistance" of space itself. This is magnetism

9. A characteristic of magnetism is the resulting field of flux surrounding the charged entity. By convention, a negatively charged entity traveling in one particular direction will have a determined orientation of its flux lines. The intensity and orientation of the flux lines are relative to the speed and direction of the charged entity. The opposite orientation of flux lines is exhibited with a positive entity with same physical parameters of speed and direction; only the orientation of flux lines has changed (for example a clockwise direction around the electron in a direction along an axis, whereas a counterclockwise direction around a proton traveling in the same direction). This realization is best described with the "field concept" as developed by Michael Faraday.

Also, and of the most significant consequences, entities of same charge and direction will attract one another (magnetic pinch effect), *repel if opposite charge with same direction!* This is a critical statement in respect to static charges whereby a static positive and negative charge attract each other and like charges repel each other.

10. **Postulate Three:** $Mo = \int_{v=0}^{v=c} v \bullet K' \bullet q$: Rest mass Mo is equal to the entity quanta, q, traveling thru space with the expansion velocity as a function of magnetic permeability Uo of space. Thus, mass is a function of the expansion energy. "q" is quanta in its "electric charge" sense, v is the expansion velocity, Uo is magnetic permeability. R^2 is the distortion of space about an idealized entity q.

- To expand this concept

- If the primordial charge, q, which in units of Coulombs, has been given energy via the Expansion of the universe as a function of the magnetic permeability Uo

- Then it followes the general form of the magnetic pinch equation of $F= (k'*q*q'*v^2)/r^2$, where $k'=u_o/4\pi$

- In which in our relative state it has now become force of attraction between charges as in law *Coulombs law*. $F = (k*q*q')/r^2$, where $k=1/(4\pi\, e_o)$

- Which forms the fundamental particle the proton with a unit charge of 1.602^{-19} *Coulombs*, which is now equated to the mass of the proton of 1.673^{-27} *kg*

- Where the K constant for Coulomb is in units of $N \bullet m^2 \bullet C^{-2}$

- The gravational constant G is in units $N \bullet m^2 \bullet kg^{-2}$, whereby the force of attraction between mass is $F = G*(m1*m2)/r^2$

- It shows that the primordial charge q has become mass in Kg

- The neutron which is exactly the same as the proton in construction but differs in mass by 2-1/2 masses of electrons which must elimanate the protons charge effect

11. **Postulate Four**: The velocity of the expansion vector is the speed of light. When this is true u_o, magnetic permeability, becomes e_o, the electric permittivity. Or at least some factor of the speed of light which depicts itself in our state of reference as the speed of light.

For example, If the universe is expanding at the velocity of light, then the magnetic force between two primordial charges q traveling side by side thru space is equivalent to *and is* the "magnetic pinch effect", -(This is the most fundamental postulate and physical process occurring in this paper) is given generally by:

$F = (k'*q*q'*v^2)/r^2$, where $k'=u_o/4\pi$

(k' is a measure of the force between charges q)

***Coulombs law is*: $F = (k*q*q')/r^2$, where $k=1/(4\pi\ e_o)$**

(k is a measure of the force between charges q)

When the expansion of the universe is equal to that of light , $v^2 = k/k'=c^2=1/(u_o*e_o)$

$(k'*q*q'*v^2)/r^2 = (k*q*q')/r^2$

Thus, the only fundamental facet of space is u_o, or magnetic permeability. When quanta travel at v=c, then the force between quanta becomes e_o or Coulombs law. The Earth, and the universe, is at a velocity of c thru space, u_o becomes e_o in our "relative state". The electric permittivity e_o is the "relative state" of u_o traveling at the expansion velocity. In many standard books in physics, The relation of the speed of light to the quantities permittivity and permeability are usually cited, but never a physical reason as to why. I offer this as a possibility to this equation. "are we bound together because of our velocity thru space-amazing if so!"

12. **Conclusions:** There is only one fundamental entity in our local universe called quanta, also loosely called corpuscles, photons, or "strings". Its mass is m=$7.37* 10^{-51}$ kg. It has an intrinsic magnetic field about it via the expansion of the universe and u_o. In our relative "stationary" state, the

attraction between quanta is e_0 or coulombs law. *They are fundamental magnetic components.* It was mentioned that in respect to static charges whereby a static positive and negative charge attract each other and like charges repel each other. The proton is an accumulation of quanta; the magnetic components bind together to a fundamental density in the shape of a sphere. There are two very real possible proton formations possible. One: Magnetic polarities of each quanta in an arrangement to make a "magnetic monopole". For example, all magnets have north and south, this is merely a relative statement for the idealized point charge/entity. For example, if they bind together in such a way as if all "south or north poles" of each quanta face outwards then the surface of the spherical proton will depict either all north or south, somewhat like an idealized monopole. Thus protons would not bind to each other, they are repelled (like charges repel), also they still have their south or north monopole field about them, thus all free quanta will still be attracted to the proton like "a sphere of uniform positive electrification", as J.J. Thomson once said. It will accumulate onto the proton until the proton field is cancelled, "an electron layers worth". Second Proton formation possibility is dipolar in aggregate which lends itself to an overall intrinsic magnetic dipole nature. Thus electron quanta layering will also be dipolar.

13 A question arises as to why is the mass of the proton much greater than the mass of the electron layer with same charge values. It is speculated that with the initial quanta formation into a sphere, the proton, it's "Guassian" field value ends up this way. To cancel or neutralize this field only requires so much quanta, an electron layers worth?

14 The nuclear force, which binds the nucleus, is a function of the neutral particle the neutron as there is no such thing as a diproton. Of course the neutron itself is an aggregation of quanta only. Whether it is a collapsed hydrogen atom with additional quanta(about 2-1/2 times an electrons

worth), or disorganized arrangement of quanta for overall charge of zero, and magnetic moment. Nevertheless, the neutron is some kind of interface between the protons.

15. The atom is an aggregation of one entity, quanta. Each quanta has an intrinsic magnetic field and is binding to the atom like a crystal lattice with vibrational modes via Planck's blackbody idealization and subsequently quantum mechanics. To break the bonds of the "layers" of the atom via n*h*f, one must match the energy which binds it to the atom. Its quantization as depicted by Planck and Einstein as being absorbed and emitted in discrete packets are the result of the fact that electron shells are the quantization of quanta shells which bind together as magnets do in discrete fundamental sizes as only so much can possibly "fit together" in discrete shells.

16. To achieve Radiation form of quanta, the energy of all rest mass or rest quanta is via the expansion velocity, its energy resides in the intrinsic magnetic field per quanta and now binding to other quanta in an atom. To break the bond, energy is needed to match the binding energy. The quanta will be broken off like a bound spring under tension. It will release isotropically from the atom in layers of n*h*f with a speed of c because this was the energy given in the beginning, and something fundamental to space itself as a limiting speed as a function of magnetic permeability. The process will continue via quantum mechanics until the plasma state or ionization state of a layer of quanta "electron" is removed, these are the electronic states, and continue linearly through ionization as n*h*f radiation still occurs as more energy is released, however now at x-ray to gamma ray wavelengths, where each period of the "wavelength" is an amount of quanta layers. A quote by Samuel Glasstone from his book "Sourcebook on Atomic Energy", "It may perhaps be permissible to consider radiation as

consisting of photons whose statistical distribution is represented by an equation of the form applicable to the propagation of waves".

17. There is no "rigid point particle" alone in space with rigid boundaries, about the entity q is the magnetic field. Its boundary walls are magnetic described by r^2", thus a "springiness" to it. It's appearance is that of a "bubble", or more precisely a "magnetic bubble". The quanta, q, has a velocity thru space via expansion vector, thus an intrinsic magnetic field about it, $B=(q*k'*v)/r^2$. the r^2 is the distortion of space about q. It is the magnetic field. The de Broglie equation of $l=h/mv$ is realized as lambda, the wavelength, is r^2 magnetic field about q. Thus, de Broglie equation is depicting the special theory of relativity whereby an increase in velocity of the mass will correspond to wavelength variations as in $B = (q*k'*v)/r^2$. An increase in velocity will change the magnetic field about q. Einstein's special theory of relativity of mass increase from $v=0$ up to $v=c$ is an energy increase via velocity thru space and exhibited in a stronger magnetic field about fundamental q as exhibited by $B= (q*k'*v)/r^2$ per quanta. ***Thus, the theory of relativity is a statement about relative velocities with respect to space, its physical significance is now understood in the realm of magnetism which in itself is a function of velocity in space.***

18. A statement by Einstein,"Thence we conclude that a balance clock at the equator must go more slowly, by a very small amount, than a similar clock situated at one of the poles under otherwise identical conditions". This is realized as the increase in energy/mass as a result of velocity increase is by the stronger attraction via the quanta within the atoms. Again the magnetic pinch effect is given by:

$F= (k'*q*q'*v^2)/r^2$, where $k'=u_o/4\pi$. The relativistic mass increase above the rest mass is realized as a stronger magnetic field within the atom themselves as a function of velocity thru space. *Thence we conclude atomic processes are affected.*

19. The relativistic mass equation $M = Mo/((1-v^2/c^2)^{1/2})$ is relative to our stationary state of the expansion velocity. The relativistic mass equation relative to "absolute space" must then be $M = Mo/((1-(v^2+c)/2c^2))^{1/2})$. The limit of velocity of free quanta is the speed of light relative to our stationary state. Relative to "absolute space" or the expansion vector, it is twice the speed of light.

20. There is no "positive" or "negative" charged entities in space. There is only one entity q with an intrinsic magnetic field as a function of its velocity thru space. In our stationary state of "now" which is at a constant speed of the value of light, they bind together as generally discussed in 12,13 above to give "relative values" of north and south about itself and consequently via coulombs law as positive and negative as Benjamin Franklin has named. Thus, we may say that the proton is a concentration of a "south field"(south or north is arbitrary, however, maybe in time there will be a proper nomenclature of south and north). An electron layers worth of quanta is attracted to the proton. An electron layers worth of quanta may be ionized from the proton, and as magnets would still bind together in "free space" as in cathode ray tubes and behave as individual particles (photoelectric effect).

21. **Gravity**: In the aggregation of magnetic quanta into atoms themselves, the overall balance that results is the attraction of gravity itself. Gravity is the resultant composite field of the atom. General relativity is the description of the curvature and distortion of space itself by mass. Quanta are the fundamental entity in the universe. It has an intrinsic magnetic field about. The magnetic field itself distorts and curves space. The whole atom is the composite field of balanced magnetic quanta. Thus, G, the

gravitational coupling between matter is the composite of u_o and e_o in the overall binding of the complete atoms. A question arises; does the "residual" effect of the Expansion Vector show itself in the ionized core of the Earth as the magnetic field? The orientation of the Earth's magnetic field is the vector of the Expansion vector, however, the spin of the Earth might "skew" the effects of actual direction.

22. *Inertia:* Lenz's law describes the "resistance of space itself" via u_o as a function of a change in velocity, only during accelerations as in all ac circuitry does it occur. Inertial mass is analogously the same during accelerations. Magnetic permeability u_o is applicable for individual quanta, free charges as electrons and protons during acceleration. It is the relativistic mass of the moving body $M = Mo/((1-v^2/c^2)^{1/2})$, and conversely $B = (q*k'*v)/r^2$. As with gravity, inertial mass is the composite of u_o and e_o of an atom during acceleration with respect to space. Inertial mass will behave analogously like ac circuitry of idealized point charges with acceleration.

23. About the entity q is the magnetic field described by r^2. This field is permanent via the expansion vector. The magnetic field is created as a result of entity q's movement thru space and space's natural resistance u_o. This is intrinsic spin.

****
*

To continue with NFG system. In reference to fig. 4 The Plasma Channel: Technically a plasma by definition is the forth state of matter and only the very center region is the "plasma zone" itself. However, I would like to encompass the name Plasma Channel to the entire region surrounding the plasma zone and the actual plasma zone itself. My reasoning for including this entire area is as follows; during plasma conduction current between electrodes, it will be evident that the plasma will have a resultant magnetic field as a result of several factors such as; established plasma current itself via conduction established by electrodes, counter magnetic fields by surrounding non plasma portion of plasma channel, hence, the field extension and coupling of energies will extend into this entire region in contact with the plasma zone itself. The gaseous and liquid zones surrounding the plasma zone will have an effect on the "resultant" magnetic fields along with many other physical properties of temperature, pressure, electromagnetic energy interchange via plasma zone and surroundings. Thus, the entire plasma channel must be viewed as a system that is dependent upon each other and cannot be separated into a discrete zone without consideration of the zones next to each other. It might also be realized that some forms of energy directly from the plasma zone will without hindrance travel right thru the gaseous zone unimpeded without any energy transfer and continue into the liquid zone. This can also extend even further in regards to some forms of energy in the gamma wavelengths which might even extend without much impedance right thru the liquid zone itself. As I stated earlier in regards to the size of the vessel to be large enough to essentially ignore the walls of the vessel to have any recoil or absorption of energy from the plasma zone. It is this inventions desire to have all energy via plasma discharge to be absorbed by surrounding fuel/water medium for complete heat transfer efficiencies. However, it is realized some forms of energies such as gamma and neutrinos might pass right thru the vessel. Hence, sufficient distance between vessel walls and plasma discharge are desired. Also, vessel walls are also constructed in such a manner to limit the amount of high energy electromagnetic energy dissipated via plasma discharge. All of these considerations have already been employed in fission vessel reactors. Thus, there is no modification required for vessel material and construction in said NFG system. It shall employ a design consistent with that of pressurized water reactors-PWR.

Let us now examine the mode of nuclear combination which is to occur in this plasma discharge. It was the scientist K.T. Bainbridge of the United states who in 1933 was the first to verify the mass-energy equation of Einstein's E=mc^2 in the reaction of hydrogen proton bombardment unto lithium with the products becoming helium

Li(7) + H(1)$\Rightarrow$ He(4) + He(4). Since this verification has taken place, numerous nuclear transformations have been studied in detail and there can be no question as to the validity of the Einstein equation or of the reality of nuclear transformations.

Besides the fusion of the so called "simplest" elements of hydrogen and its isotopes, there is a more fundamental step in which I believe is being ignored and could possibly be the exact very process occurring in "Cold Fusion" or CF experiments which give anomalous effects of positive net energy generation. Though CF is not accepted by the vast majority of plasma scientists. It has a question mark as to its possibilities as does superconductivity possibilities at room temperature. Nevertheless, I believe this said process I am putting forward at its lowest energy considerations just might be the actual cause of Cold Fusion occurring in an electrolytic fashion. I believe the most fundamental particle combination for nuclear combination is that of a proton and an electron itself. It is known that the beta decay of a free neutron is approximately 20 minutes. Thus , there is no such thing as a free neutron as there is a free proton. And a free neutron is in fact the simplest radioactive species. If a free neutron has transpired via artificial or natural means such as radioactivity, it is known that the decay of a neutron will produce a proton and a electron and approximately 0.782 mev of energy in an exothermic reaction. This energy is divided between the ejected electron and a neutrino. 0.782mev is also 1-1/2 times the electron rest mass, also, since one electron was also produced, it is understood that the neutron seems to be at least 2-1/2 times an electron mass greater than a proton. Under these considerations it is my belief that it is possible to fuse electrons upon a proton and create a neutron. The minimum energy anticipated is 0.782mev. Once a neutron is created it is quickly "absorbed" by another free proton into a stable element call a deuteron. This neutron capture process is well known and described as a Radiative capture type.

(H + n $\Rightarrow$ D +γ where γ is a gamma ray photon of 2.2.mev)

This reaction is exothermic with approximately 2.2mev emitted as a gamma ray photon. Hence the surplus of energy derived 2.2mev minus 0.782mev equals 1.141mev.

It might seem apparent that the gamma ray photon emitted is somewhat useless because of its extremely high penetrating power and as such would be ineffective at transferring its energy into a thermal range for extraction by said NFG system. However, I believe because this system is entirely filled with a liquid hydrogenous solution the rate of absorption via photoelectric effect, Compton scattering, and Bremstrahlung that most of the gamma photon will be absorbed.

Let us now look closer at the actual plasma state created in said system. It should be realized by now that what is being presented is simply an opposing electrode assembly in which a large electrostatic potential will be applied to each electrode and cause a dielectric breakdown in liquid water to occur. The characteristics of the plasma formed are an impedance load as opposed to being purely resistive. Thus, it is clear that with an alternating current being used a resonant frequency be employed to properly match plasma state. Also this system is again based upon the sole focus of magnetic energy, using a higher frequency not only depicts itself as higher energy in Planck's equation of E=hf, but a high frequency will also increase the magnetic component per oscillation of electric pulse. This is realized as the fact that as the frequency is increased, the wavelength decreases, the entire envelope of energy is in a smaller dimension. Light photons are smaller that microwave photons. X-rays even smaller. It seems that with higher frequencies, the actual envelope of energy is within a smaller package. This is somewhat related to the fact that a magnetic field is created as a function of velocity as shown earlier as

B=k*q*v , if the rate of change of alternations is greater, then the velocity must be greater as known from simple understanding of sinusoidal waves. For a wave to oscillate quicker, its rise and fall times must be steeper, hence, velocity is greater. This is also deduced from Faradays law also in regards to rate of change in current or magnetic flux. However, I would like to point out that this is also a statement in relativistic mechanics. The theory of relativity describes the change in mass, which is energy, of a body with velocity increase as depicted by the

"relativistic mass equation" $M= Mo/((1-v^2/c^2)^{1/2})$.

Conversely the debroglie equation shows a wavelength character for mass with velocity.

$\lambda=h/mv$, deBroglie wavelength

The masses in our case are electrons and protons subjected to large voltages and made to move with a velocity. It is these velocity factors which are impinged upon the particles in said plasma at such a proper range of frequency that a packet of magnetic energy of sufficient flux density and "quantum size" will be generated as to create the favorable environment for fusion of said nuclear material.

Now that we have deduced our reasoning for the use of alternating current of high frequency, the question as what to what resonant frequency would be "most efficient". Immediately the 21cm or 1.420 GHz comes to mind which is related to both spins of the 1s electron and the proton in hydrogen, however, the frequency might be dependent upon many physical parameters such as fuel type, pressure, electrode spacing, etc In regards to the power source which will drive the high voltage and high frequency to said electrodes, it was noted that the preference for a Tesla coil is desired for personal reasons by the author as I share a great respect for Dr. Tesla and his strong interests to provide unlimited energy for all humanity. It is for this sole reason alone in that it would be sweet justice to have the marriage between a Tesla coil and my NFG system to fulfill our similar interests in providing energy to everyone efficiently and affordable. A quote by Tesla "Of the various branches of electrical investigation, perhaps the most interesting and immediately the most promising is that dealing with alternating currents. The progress in this branch of applied science has been so great in recent years that it justifies the most sanguine hopes. Hardly have we become familiar with one fact, when novel experiences are met with and new avenues of research are opened. Even at this hour possibilities not dreamed of before are, by the use of these currents, partly realized. As In nature all is ebb and tide, all is wave motion, so it seems that in all branches of industry alternating currents— electric wave motion—will have the sway". However, it should be clear that any form of power source which may produce said optimum electrical parameters of high voltage and high frequency will be suitable. Let me continue the description of NFG system with the Tesla coil in mind. The

Tesla coil is a system that has the ability to resonant an alternating current between its inductive and capacitive coils. In doing so it has the distinctive behavior of being able to produce very large voltages in the range of millions with high frequencies usually in the kHz range. To change the resonant frequency of the Tesla coil, parameters of inductance and capacitance of the coil itself are changed to produce a somewhat fixed output in frequency. As was noted earlier that there exists the possibility of "frequency drift" in plasma state of said system by various parameters as pressure, temperature and magnetic fields, The Tesla coil (or appropriate power source) should have the ability to be adjustable being some kind of "phase lock loop" feedback signal. A feedback system has the ability to monitor output conditions (plasma channel in this case) and vary the input signal accordingly for proper power matching conditions between input and output.

In regards to the plasma zone itself. This area is made up of free particles of hydrogen and oxygen as the free positive ions, and free negative electrons in presently described system with only pure water as the fuel. It was stated earlier that the velocity of a charged particle in an electric field is equal to

$$v = \sqrt{(2qV)/m} \quad \text{(equation 4)}$$

also, the force of the magnetic pinch effect was derived as

$$F = k'q1q2\frac{v^2}{r^2} \quad \text{(equation 8)} \quad F = k'q1q2\frac{v^2}{r^2} = kq1q2/r^2$$

(equation 9)

Thus by equating the magnetic pinch equation to that of Coulombs law, I deduce that relativistic speeds of charged particles must occur for magnetic pinching and fusion, specifically a value equal to the speed of light in magnitude has a significant meaning. Therefore, as a starting point I use the speed of light as a magnitude for the required velocity required by the charged particle, I can obtain a voltage requirement used for its velocity from equation(5)

$$\text{velocity of charged particle} = v = \sqrt{(2qV)/m} \quad \text{(equation 4)}$$

$$\text{Voltage} = V = v^2\frac{m}{2q}$$

Theoretical voltage required for electron to approach speed of light = ((v^2 * mass) / (2*q)) or ((9x10^16m/s)*(9.11^-31kg)/(2*1.6^-19coulombs))= 256Kilovolts

A voltage of 256kv is initially thought to be used to accelerate an electron to the speed of light in a vacuum. However, since the early days of cathode rays tubes and beta ray studies, a phenomenon of e/m (the ratio of electron charge to its mass) was variable with increasing speeds. It was H.A. Lorentz

who derived an expression $\sqrt{1 - \dfrac{v^2}{c^2}}$ which was earlier proposed by G.F.

Fitzgerald in trying to resolve the stationary ether paradox of the Michelson Morley experiment. In further studies by Einstein using the principles of H.A. Lorentz concluded that any particle will behave according to the relativistic mass equation whereby, $M = Mo/((1-v^2/c^2)^{1/2})$. Hence, the mass of the electron will not reach the speed of light in our "relative state", however, its mass will be greater as we try to increase speed. Thus the voltage requirement of 256kv will not suffice and a much larger voltage will be required for relativistic speeds. Approaching the speed of light the mass of the electron can reach many times its original mass. By studying the relative mass equation a close approximation to near relativistic speeds, it is deduced that a value of 10 (ten) times the original rest mass of the electron (or any particle) may be used as a good approximation. Therefore, 10 (ten) times the rest mass is used and accordingly the voltage will increase by ten times as a requirement for reaching the relativistic speeds. This will make our original voltage requirement of 256kv become 2.56million volts as the necessary voltage for velocity attainment of electrons.

This voltage is impressed upon the entire plasma current. The arc length between electrodes will be a function of reactor size, thus, to provide the amount of energy for charged particles distributed about the plasma arc a greater voltage is used across electrodes themselves and accordingly the proper amount of voltage is distributed across plasma arc length to initiate fusion conditions. Let us merely state one hundred times calculated voltage is required at this point between electrodes with a reactor size conducive for electrode spacing at one meter apart. Thus, 256 million volts has become 256 million volts with a plasma arc length of one meter being employed.

Since an extremely high voltage is required for said relativistic speeds, the need for a high frequency is required to establish this high voltage. It was noted earlier that the frequency of this impedance load must be made to resonant with applied voltage input source. This should be realized as with all alternating systems and impedance loads. Thus, it should be clear that it is imperative to first establish a very high voltage for plasma conduction at relativistic speeds of some particles in plasma zone, and concurrently apply a high enough frequency for several reasons as already stated. The transition from a low frequency current to that of a higher frequency has some peculiar effects such as skin effect. Skin effect has the characteristic property of forcing the current onto the surface of the conducting wire. These effects are related via the magnetic component of alternating current. The well-known equation for skin depth is given below. Note that skin depth

(δ) is a function of only three variables, frequency (f), resistivity (P), and relative permeability (μ_R).

$$Skin\ Depth = \delta_s = \sqrt{\frac{2\rho}{2\pi f \mu_0 \mu_R}}$$

where :
$\rho = bulk\ resitivity\ (ohm - meters)$
$f = frequency\ (Hertz)$
$\mu_0 = permeability\ constant\ (Henries\ /\ meter) = 4\pi \times 10^{-7}$
$\mu_r = relative\ permeability\ (usually \sim 1)$

The important point to see here is that the magnetic permeability has a direct bearing on skin effect as a function of frequency. It has been stated that the magnetic field is a function of a charged particles velocity in space as in equation five

$B = k'qvX\, r/\, r^2$, (equation 5) units in S.I. are Tesla for magnetic flux density

If alternating current is used and applied to electrons in a conducting wire, or a plasma current, the charged particles will be subjected to an oscillating electric field and hence with a greater oscillation, the velocity effects must be greater, and thereby the magnetic fields will alter.

In furthering this action of higher frequencies and high voltages the reader is asked to keep in mind the output plasma arc of a Tesla coil. The rf current from coil outputs see's its path to ground as one of impedance. The high voltage establishes the dielectric breakdown associated in air when such Tesla coil is operated. There cannot be one without the other. The high voltage must be established for plasma conduction and a high frequency is required for additional magnetic qualities. The transition from low frequency effects to that of high frequency is subtle in the very least. When we operate a dipole antennae with an appropriate oscillating current, we think in terms of quarter wavelengths and utilize proper matching techniques to impedance match the incoming signal and "launch the wave" onto some kind of conducting wire acting as an antenna with dimensions associated with wavelength. When we say for example we have a 100mhz signal, we say the electric peak to peak signal is 3 meters, where wavelength is $\lambda = c / f$ in a vacuum, do we think how wide the width of the magnetic effects are. To continue these thoughts, we think of a magnetron and its standard 2.45ghz frequency output. It has an electric peak to peak wavelength of 12 cm. Does the magnetic field also coincide to this value in width as it represents the perpendicular value to the electric field. Is this "wave" symmetrical as it is launched from the cavity in regards to a packet of energy with not only a peak to peak electric wavelength, but also a perpendicular dimension as a higher frequency is attained. To continue this thought, EM waves of light frequency magnitudes which are in the nanometers are used in optical microscopes for resolution and determine resolutions based upon wavelength considerations. We can differentiate between molecular dimensions in optical dimensions because we rely on the fact of the so called peak to peak wavelength resolution which we must agree has dimensions that are directly related to not only peak to peak for the electric magnitude, but also the magnetic. Though it is known there does lie the relationship between the electric field and the magnetic field of electromagnetic waves in free space as in Maxwell equations, I am referring to "magnetic peak to peak" dimension as it coincides with its counterpart the "electric peak to peak" wavelength. For in this plasma arc being formed in said NFG system, we are not launching a wave at the input frequency, we are oscillating charges at this frequency, and in doing so, peculiar effects of magnetism are associated with it. I believe the magnetic field component which is always relative to the electric field component as in Maxwell's laws as we progress into higher frequencies the adjoining peak magnetic components began to attract to one another much more tightly. The higher the frequency, the closer are the peaks to one another. It is the magnetic components being squeezed together by the action of the higher frequencies which makes the resultant packet of energy hold together much more than at low frequencies. For example, the magnetron output in a standard microwave oven has a wavelength of 12 cm, which may also be divided into two packets of energy of opposite electric and magnetic field values. Each packet size must be approximately 6cm. This packet of energy can bounce off the interior walls of the oven and will not penetrate the glass door. This glass door has a mesh screen with holes that should be much less than ¼ wavelength and accordingly it is. The packet of energy does not escape. However, in studying the duality nature of electromagnetic waves, some experiments such as by Thomas Young as early as 1800 have given more credit to waves as opposed to corpuscles because of the effect of a wave front on a surface with some of the wave passing thru an aperture or slit in it and the resulting patterns of interference formed. Of course the slits in those experiments are far greater than the wavelength of light being in nanometers. I specifically am referring to the fact that slits or holes in mesh screens smaller than ¼ wavelength of a particular wave front do not allow any appreciable amount of the "wave front" to escape. As the frequency gets higher, the packet of energy is held together tighter. I referred to this earlier as with an increase in alternating current, the angular velocity must increase accordingly, hence, magnetic effects are increased. It seems that a 'quantum" of energy is increasingly held together as the frequency of oscillation is raised. The quantified 6cm packet of energy hits the mesh screen and does not let any part of its "quanta-photons" escape as it is held tightly together. To illustrate another step further in velocity considerations, for an increase in resolution in microscopes beyond the light regions, we advance into the scanning electron microscopes whereby the electron has a "wavelike" behavior via the deBroglie equation as a direct function of its velocity in space

$\lambda = h/mv$, deBroglie wavelength (equation 9)

A high voltage is used to obtain large speeds attained by the electron, a wavelike picture of its dimensions should be estimated via the deBroglie equation of $\lambda = h/mv$ whereby lambda λ or wavelength is a function of the Planck constant divided by the product of momentum mv. Thus it should be clear there is another dimension of "magnetic wavelengths" which must be taken into consideration when in said method of relativistic speeds is required for magnetic pinch qualities. With the values established for relativistic mass of the electron (10 times original electron mass times velocity of light) we can somewhat come close to a value of lambda. It is conceded, that the actual values of mass and velocity are not exactly that given, but should be a close approximation when applied voltages in the range of several billion are used. Hence, a value of 2.4^{-13}meters for a deBroglie wavelength of an electron is found. This is close to nuclear dimensions of the proton. Thus a physical picture of the electron should now be established. An electron given relativistic speeds in said NFG system will acquire a stronger magnetic field about itself and its dimensions begin to shrink because of the strength of its magnetic field compressing it. This ideology is also understood in cross section neutron studies whereby the energetic neutrons with varying velocities posses different wave particle duality conditions of wavelengths and entire encyclopedias have been derived for cross sections studies of neutrons and target elements. Hence, the electron in our case has been reduced in size via description of its magnetic field in a way which can have a direct relationship in reacting with nuclei as in neutron cross section studies.

Therefore, we have established in the plasma zone an amount of relativistic electrons having cross sectional dimensions on the order of nuclear sizes in which interaction between said electron and proton can fuse via magnetic attraction. It was stated that the radioactive free neutron will spontaneously emit an electron and 0.782mev of energy. Conversely, if an element has too many protons relative to neutrons, the nucleus will create an inverse beta decay or k-capture or also called orbital electron capture. Thus, it is evident that this proton electron combination process seems plausible. Therefore, in analyzing our plasma zone, in a "snapshot instant in time" when the magnetic energy for pinching is highest within the plasma zone is a relativistic electron traveling in a direction towards the positive polarity of positive electrode. This electron will have a magnetic field created by its velocity and its orientation in regards to north and south may be established via known methods of field conventions. Conversely, at the very moment this "snapshot instant in time" is taken, positive ions of H, D, T, and oxygen will move in opposite direction of electrons toward negative potential of electrode. Its associated magnetic field strength is different than the electron because of its mass being much greater than the electron and subsequently with a given voltage applied, will be forced to move with a velocity less than the electron by way of equation 4. Mass for proton is much greater. Also if the mass is a deuteron, the velocity will be even less as a neutron is attached and the appropriate mass is doubled in equation 4.

Velocity of charged particle = $v = \sqrt{(2qV)/m}$ (equation 4)

However, more importantly, its magnetic field by convention is opposite than the electron. Though the orientation of the magnetic field lines of the positive ions are opposite to that of the negative electron, because its direction is opposite to that of the electron, the fields will be the same. Thus, the magnetic pinch effect not only is attractive to all negative electron current in one direction, but also is attractive to positive ions traversing in opposite direction within plasma zone. Thus, the magnetic pinch effect is additive to entire charged particles within plasma zone as shown in figure 5. This of course is true in any plasma current pinch device.

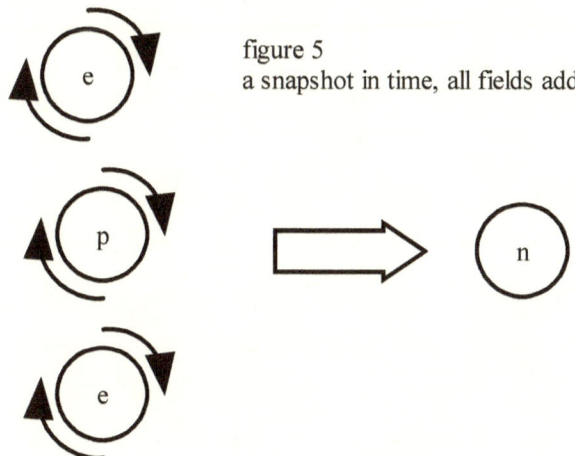

figure 5
a snapshot in time, all fields add

Therefore, in this instant in time of the snapshot, there lies the probability of a proton being surrounded by the requisite number of electrons needed for neutron formation. I have repeated the need for relativistic charged particles to acquire a velocity component which depicts itself as the required magnetic field needed for fusion pinch, to state it another way, the velocity component is a direct result of the voltage component impressed upon the charged particles. I have referred to the fact that at least 750kiloelectronvolts of energy is required in the vicinity of charged particles for a "reverse beta decay" and subsequent neutron formation to occur. When I stated values of several billion volts impressed upon the plasma current, I am trying to convey the fact that this distribution of energy must be allocated in the strength of the electric field amongst charged particles. Hence, once a free neutron is created in a hydrogenous solution such as water it will immediately be absorbed as hydrogen's thermal neutron cross section is extremely large in comparison to D,T,and Oxygen. Thus a deuteron is formed. I shall refer to this mode of deuteron formation via initial neutron formation "case No.1".

Case no.1 = (H + n $\Rightarrow$ D +γ where γ is a gamma ray photon of 2.2.mev)

It is conceded that with initial neutron formation, its velocity is not guaranteed to be thermalized; however, it should be evident that amongst the material elements being employed in said NFG system, the absorption by hydrogen is far greater than any other element. The following table is a comparison of some elements FROM "SLOWING DOWN OF 1-MEV NEUTRONS" TAKEN FROM SECTION 11.40 FROM "SOURCEBOOK ON ATOMIC ENERGY" BY Samuel Glasstone, 2nd edition

ELEMENT	H	D	O
MASS NUMBER	1	2	16
FRACTIONAL ENERGY LOSS PER COLLISION	0.63	0.52	0.11
COLLISIONS FOR THERMALIZATION	18	25	150

• CAPTURE CROSS SECTION (BARNS)	• 0. 33	• 0.000 46	• 0.00 02

This case no.1 represents the lowest level of a fusion reaction. This case no. 1 is also an understanding into "cold fusion" reactions, whereby an electrical current in a hydrogenous solution can provide a magnetic environment such as shown in figure 5 and create the necessary conditions for a fusion reaction. Cold fusion reactions are not accepted as there has not been definitive proof of excess energy, by understanding the present modes of magnetic excitation, it should be understood that cold fusion cells employed are at the lowest energy realm possible for fusion. The intermittent positive values detected are via quantum tunneling and probability analysis with modes of excitation and fusion combination as I have stated and not at all via theories explained by cold fusion proponents such as a function of electrode material by catalytic actions. The need for a higher voltage and higher frequency must be employed for optimum conditions. Nuclear reactions as with any chemical reactions are somewhat identical in the fact that for the most efficient transfer of input energy into the system for reaction to occur multiple parameters must be optimized. Reactions are similar to a bell curve in that the peak of such a reaction curve represents the proper conditions of all parameters. In this case, voltage, frequency, pressure, electrode spacing, etc are all important. However, some reactions will still occur on the very leading point of the efficiency curve, this is where cold fusion is occurring.

Refer to figure 6. On considering the next higher level of a fusion reaction in such NFG system, we again analyze the same snapshot in time. It should be realized that there also lies the probability that at the very instant of this "snapshot in time", when the magnetic energy is most effective for pinching, that there can exist two adjacent protons with a proper amount of electrons in its vicinity to cause a direct deuteron formation. However, the electrons in this case only facilitate the pushing of one proton onto another via its magnetic energy in a direct hydrogen-hydrogen fusion reaction.

Case no. 2 = (H + H $\Rightarrow$ D + eplus + energy, note, the eplus is a positron)

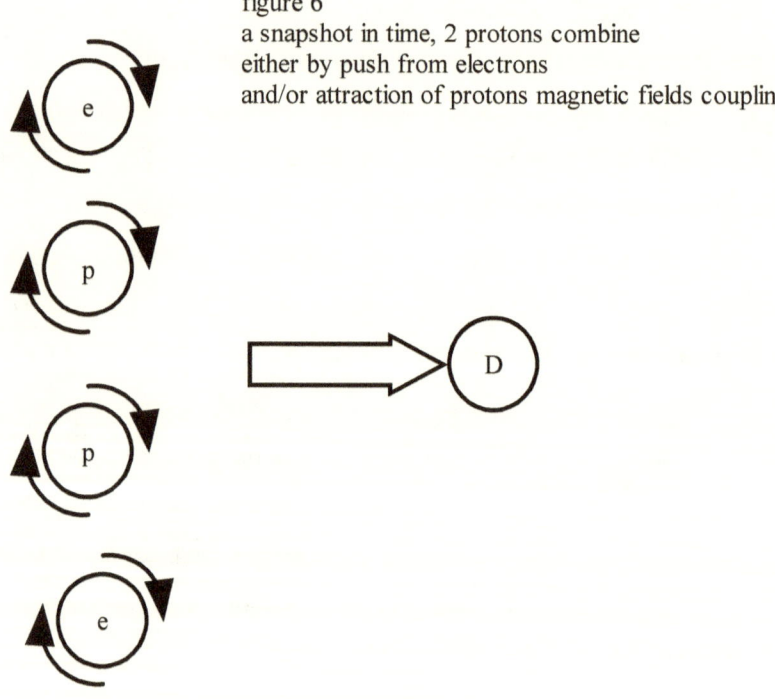

figure 6
a snapshot in time, 2 protons combine
either by push from electrons
and/or attraction of protons magnetic fields coupling

The following represent levels of fusion reactions which may exist in every opportunity of a "snapshot in time". Whether the reaction is by direct combination of electrons and protons as described by case no. 1 or by the electrons causing nuclei to fuse as noted in case 2. Fusion caused by the relativistic magnetic field of the electrons aiding in pinching the overall field between electron and protons when all entities are in proper positions, or by the momentum of the relativistic electrons causing a direct collision with a proton and slightly changing the protons course into another adjacent proton for fusion into deuteron. It should become clear that there exits many modes of fusion of elements under this method of excitement.

Case no. 3 = (H + D $\Rightarrow$ He3 + energy)

Case no. Z = (element x + element y $\Rightarrow$ element xy + energy) Equation no.13

whereby element xy is a new element caused by the fusion of elements x and y to yield a new element xy which is less than the total binding energy of the strongest element that being of nickel-62 and or iron-58/56. Thus, case no. Z represents the fusion of any material for useful nuclear fusion energy production, as long as the product element xy is an element less than that of nickel and or iron. Reference to the packing fraction by F.W. Aston and subsequent binding energy curves suggest fusion possibilities of the general equation no.13 as shown in case. No. Z.

However, it is known that fusion of the lightest elements, particularly hydrogen and its isotopes, seems most plausible and likely to occur in man made fusion devices to date because of its single bound positive charge proton. In this NFG system, it may be attainable to attempt equation13 by the deployment of concentrations of any element x and y in the medium of water thereby introducing in that "snapshot instant in time" analysis a probability of element x and element y to be in the right place at the right time for fusion to occur.

It has been portrayed that to attain fusion by said NFG system, the requirements of a high voltage are needed in conjunction with a high frequency. The imparted velocities of some charged particles will reach relativistic speeds, and in doing so attain a greater magnetic field, conversely, with a high frequency being employed for not only resonant matching techniques, but also in regards to velocity inducement as it was mentioned earlier with the fact that a higher frequency induces a magnetic quality to the oscillation necessary in this mode of application. This relates to skin effect, whereby it is known that a high frequency current will not penetrate a conducting surface appreciable. The plasma current is also a conductor; however, the skin depths desired with such NFG system are only several atoms thick. Thus, this high voltage high frequency current will have its actions about the plasma in such a way as to directly interact at the atomic level much more directly than with a low frequency current. By use of the terms high and low frequency of power line usage, one must ask what is high or low, this are relative statements. However, it should be realized that high frequency in this system is comparable to those which excite nuclear energies. It was already stated of the possible resonant frequency of the 21cm line. However, it is realized as mentioned before, multiple parameters are included and one of which may also be electrode spacing in which shall be discussed now.

In the following discussion of electrodes, the reader is asked to keep in mind the discharge of a tesla coil. The output of such a device is variable dependent upon construction techniques and proper matching of simple resonant components such as inductor and capacitor to form a resonant oscillating system. In regards to what has been discussed so far, the reader is asked to accept a tesla coil having been built with the following output characteristics; a voltage to ground of 256 million, and a frequency of 1.420 GHz or 21cm. Call this Tesla coil "TC1". The entire TC1 construction is immaterial at this point, the only requirement is that the typical "top load" of said TC1 is one of the electrodes in the NFG system. In reference to fig.7, it is shown that the electrode assembly introduced in fig 2 earlier has been replaced by a spherical design made of conducting metal of the "doorknob" type in shape. It has been found that the discharge from high frequency high voltage discharges prefer rounded electrode discharge geometry, hence, at this stage preference is made to a metal electrode of spherical design. Electrode one represents the actual discharge electrode of a Tesla coil, whether the actual power source of the coil is in the vessel itself or external with coil outputs to discharge electrode no. 1 is irrelevant at this point. Concern should be directed only to electrodes at this

Figure 7. discharge electrodes in fluid medium of water

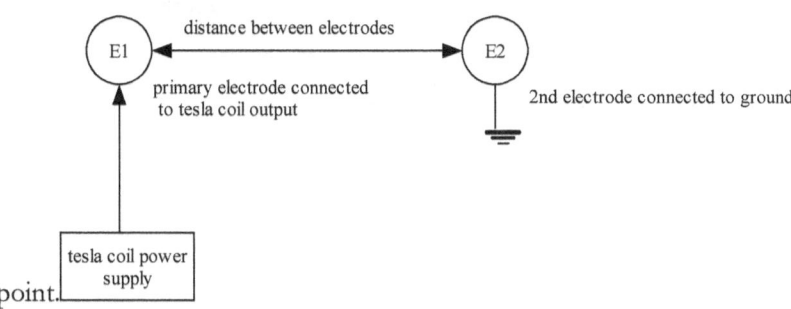

point.

At present, there are two electrodes being shown. Electrode no.1 is the Primary discharge electrode of the tesla coil. Electrode no.2 is the RF ground connection in which the RF current from electrode no. 1 may return to ground through the dielectric medium of water in this case. Thus, what has now been established is the simplest case of an electrode assembly within the NFG system which employs one electrode being connected to the tesla coil. The output of the tesla coil is discharged from electrode no.1 to electrode no.2 which is connected to a solid earth ground. The size and shape of electrodes will have a direct bearing on the impedance parameters of the electrodes as seen by the power source. For example, as with all power supplies, the source of energy doesn't actually care what number of components it is connected to, but only on the final resultant impedance. For example, the tesla coil output from discharge to ground may travel thru any medium whatsoever not just free air, however, the parameters of resistance and impedance are the only variables in which the power source must address. Thus for efficiencies, the overall impedance is not only a function of electrode shape and size, but also the plasma conditions in between.

A second design may also be used in the electrode assembly as in figure 8. Instead of a second electrode used as RF ground, it is seen that the vessel walls themselves may be used as the discharge return to ground. Thus, a single electrode may suffice within the vessel as long as it is the discharge electrode of the Tesla coil.

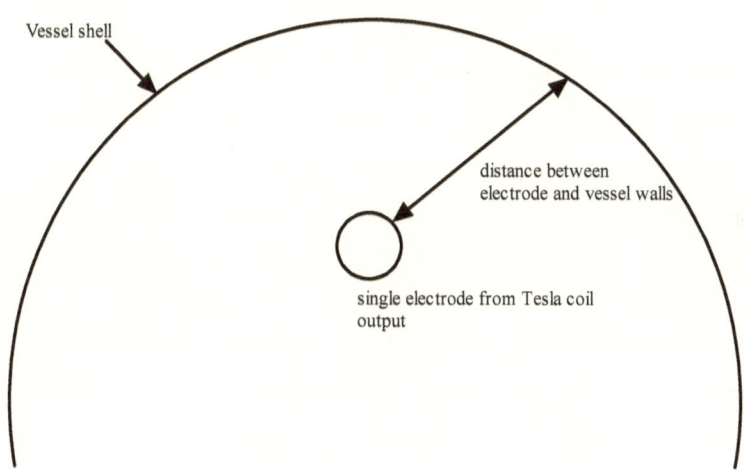

figure 8. single electrode from output of Tesla coil using shell of vessel as rf ground.

Vessel shell

distance between electrode and vessel walls

single electrode from Tesla coil output

A third design may also be used in the electrode assembly whereby the primary discharge electrode is in the center of the vessel with a plurality of secondary electrodes of rf ground spherical arranged around the primary electrode about a single axis to divide the plasma current into several segments.

A forth design may also be used in the electrode assembly whereby the primary discharge electrode is in the center of the vessel with a plurality of secondary electrodes of rf ground spherical arranged around the primary electrode about multiple axis.

A fifth design may also be used in the electrode assembly whereby the primary discharge electrode within the vessel is discharged to another primary discharge electrode of a second tesla coil such as a dipolar tesla coil. This arrangement is seen as two identical tesla coils with same construction parameters and fundamental frequencies are alike and which are made to "resonant" with one another.

It should be clear that by mentioning various electrode types, quantities and spacing about one another, an unlimited geometry of electrodes may be created within a vessel as might be required for efficiency of plasma operation and its thermal division of power translated into fluid medium

Let us now return to the simplest case of two opposing electrodes with some distance to one another. In the simplest of designs, the distance between electrodes is fixed to one another. It may be found that for a given size NFG system of a certain power rating, a determined fixed distance from one electrode to the other shall suffice. However In some instances it might be required for one electrode to me made movable in that this movable electrode is moved to a close proximity of the discharge electrode for a plasma arc to be initiated. Once an arc has been established, this movable electrode is made to move away from discharge electrode whereby lengthening the plasma arc itself. In order for plasma arc not to extinguish, the power input is increased in a linear fashion as the distance between electrodes is increased. This situation may be required if a very long plasma arc is required for whatever reasons. One reason shall be noted now; it was stated earlier in the necessity of establishing a magnetic pinch environment for the actual cause of nuclear fusion. It was mentioned that there lies the possibility of introducing external magnification to add in the total magnetic energy required. Therefore, with reference to fig 9, a source of magnification shall be introduced about the plasma current itself. This source of external magnification may be electromagnetic or fixed permanent magnets.

Fig 9. Depiction of external magnetic field about the plasma current between electrodes . This external field shall aid in pinching of plasma

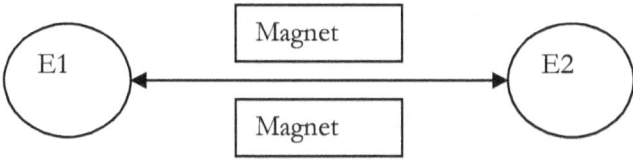

Now that it has been stated the possibility of increasing the fusion yield by employing external magnification upon the plasma arc, it might be asked why operate such a NFG system without the highest possible fusion yields in which all parameters of input such as voltage , frequency, and even pressure applied to overall liquid state are highly utilized. The attainment of commercial nuclear fusion is highly desired for obvious results. However, the use of fusion devices have always been envisioned for the use of electrical power conversion systems. Though that is exactly what the NFG system is, it should be understood that the commercial construction of an electrical power generation unit of this design, its energy output must be at least 4 times greater than input. For example, it is known that to extract energy from a thermodynamic system such as from a liquid state, approximately 30 percent can only be achieved because of the laws of thermodynamics. Thus, if we inject 1megawatt of electricity into NFG system, we should at least generate 4 times as much energy into the liquid so that enough energy is extracted from the vessel to not only provide energy surplus, but also replenish what was input into the system. That point should be obvious. However, in regards to why not operate the NFG system at a much lower potential, here should make the case. The NFG system is nothing more than a "heat source". Oil, gas, coal, and even electricity is used as a heat source for commercial buildings in heat exchange systems such as in boilers in whereby the heated liquid state is used as a transfer mechanism to supply warmth to the building. Operating the NFG system at a much lower efficiency for use in such boiler apparatuses seems plausible for cost effectiveness in a much less robust design as compared to an electrical power generation unit. For example, a boiler system employing the NFG system might be a small compact unit identical to standard boiler systems of today. The use of higher pressure might not be used for safety purposes, many parameters of input control may severely be degraded with a boiler NFG system. The bottom line will still be a much more efficient method of having a "heat source" employed, and thus, the use of oil, gas, coal, may be substituted by employing the NFG system in boilers. This should dramatically alone reduce the greenhouse gas emissions, and in so doing should be implemented immediately as an alternative to heating buildings. Of course, once the efficiency of the NFG system is made so great as in electrical power generation systems, electricity will become so cheap and abundant, that all buildings will then simply use electric heating solely from a utility company.

In reference to frequency, it was noted that 1.45 GHz or the 21 cm wavelength has significance. Let us now incorporate this frequency to a Nuclear Fusion Generator machine with following characteristics.

Voltage: 256 million volts

Frequency: 1.420 GHz

Electrode spacing: 1 meter   , thus, close to 5 oscillations are between the electrodes, conditions for standing wave ratios are sought.

figure 10. electrodes spaced at 1 meter apart, frequency is 1.420ghz approximately 5 oscillations between electrodes

strong E field creates relativistic charges with magnetic fields able to fuse nuclei in immediate vicinity

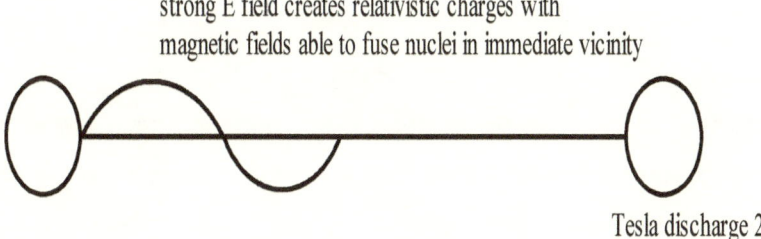

Tesla discharge 2

Tesla discharge one

In summary, it has been stated the fusion in this NFG system is by magnetic fusion alone. Without question, the magnetic energy is derived by the relativistic effects of velocity by charged particles. However, it was mentioned that frequency is also important for resonance and impedance matching techniques upon the plasma state. This seems logical as most would make analogies to ac circuitry. However, it was also mentioned, that there are inherent magnetic effects with a high frequency. Starting from low frequency effects and approaching the higher modes of oscillation we interact with the nucleus itself with high level x-rays and gamma rays strictly reserved for nuclear transitions. May this relate to the debroglie equation of matter waves $\lambda=h/mv$, whereby velocity effects might be related to number of oscillations per second in frequency response, and or also to the Planck equation of $E=nhf$. There seem many subtle parameters all of which must be related, and not simply a large discharge of electrical current from one electrode to another and simply summing the poloidal current as in standard tokamaks and similar pinch devices. What is presented here is an entirely different mode of application in fusion. A high voltage and high frequency used conjointly to impress a magnetic strength of not only quantity in flux expressed in Tesla units, but also in quality as in regards to cross section wave picture which is a function of frequency and applied voltage together.

Let me try to give another analogy to make the point. think about striking a match, if all conditions are proper , one knows that to strike a match, you must go a minimum speed- you cannot strike the match too slowly- this is understood as more speed is more friction and thus more activation energy necessary to create combustion of the match material. The same applies to this fusion system; first, imagine my system as inside a pressurized water reactor used in fission plants- the moderator is already heavy water- we shall use it as the fuel. I have said the plasma arc looks exactly like our friend the electric lightning bolt- but i talk about in my pdf file- that the so called lightning bolt must be understood in regards to high voltage and frequency- just like the minimum speed needed for the match- so it is with the combination on high voltage and frequency to not only resonate with the dielectric molecule of heavy water- but more so- to create a minimum velocity upon charged particles in the fuel water. this velocity is also exhibiting itself as a magnetic field-think of the cathode ray scope- in a snap shot instant in time for analysis, look at when the maximum energy is being applied on a per strike basis of the input cycle- or shall we say the highest amplitude of the ac signal. if for example- an input energy of one megawattt was injected into the strike of the arc- a magnetic pinch shall be directed upon the charged particles in transit of the discharge- in a thermodynamic extraction process such as this - we need at least 5 times more energy out of the system to recoup our initial energy input- and a surplus for commercial energy supply- thus- via fusion of hydrogen to helium- with each strike- a minimum voltage and frequency SHALL give the minimum activation energy required for magnetic pinching and fusion of the isotopes of hydrogen. to increase the q of the reaction in this system- a magnetic toroid may be used around the plasma arc for increased efficiency= such as used in tokamak devices. I hope this analogy helps those who question this system-

Here is a thought experiment with Einstein in mind- you are a water molecule-and you live in a lightning bolt- i propose electromagnetic fusion to end the energy crisis

Pretend you are water-you are an oxygen atom- you are stable- you are noble- 8 protons-8 neutrons-and 8 electrons- there are only 5 magic shell nuclear elements of the periodic table- oxygen is one of them- it is very stable-

Therefore- you are stable and noble- now you are this noble one and you have outstretched in your hands a hydrogen atom and or its isotope deuteron and you hold it out to mankind as a gift- a water molecule

Man understands the gift of fusion for many years and is desperately trying to do such-

I propose electromagnetic fusion with heavy water to end the energy crisis- I propose the BENNET pinch used in poloidal currents used in plasmas of tokomaks since the 50's- but much further-

Here we go – the thought experiment- you are this oxygen atom with two hydrogen's and you stand amongst your friends similar in nature. You are placed in a large vessel filled of your kind-now imagine that two walls opposing in this vessel are the plates of a capacitor- who cares what is applied to the capacitor plates (electrodes of system)-for all you know as a noble oxygen and the hydrogen you have in your hands is nothing more than the electric field upon the plates-

You are composed of charged particles- thus you will interact with the applied electric field of the plates-this is common electronics and electric knowledge-

However, I speak of dielectric breakdown- I speak of the lightning bolt- let us now assume we have made the capacitor plates oscillate at 1 million volts peak to peak as way of a tesla coil

Think of the electric field- everything will be controlled by this field- a dielectric breakdown will occur- and all discharge current will begin to flow and oscillate as a function of the applied voltage-

Now- you are in the heart of a lightning bolt- you who are noble as a oxygen probably lost all your valence electrons due to the magnitude of such a high electric field- every charged particle in transit of the discharge current is surely ionized and talks of being a complete water molecule should be erased- the state of this plasma current is nothing more than ionized hydrogen and oxygen and a complete sea of electrons-

Let us think first of the electrons- 1896 times smaller than protons- no question it will oscillate much faster as a function to the applied voltage- remember the cathode ray scope- basic physics also will say this charged particle will also have a magnetic field about itself because of the electric field that has driven it-thus- this oscillating electron current will have an intrinsic magnetic field

Let us now think of the protons- oxygen has 8 and surrounded by 8 neutrons and is far less likely for reaction than isotopes of hydrogen for fusion- this should merely be understood by refereeing to atomic tables of elements and known theory- but this hydrogen is a singly bound proton with mass one- unless we speak of a deuteron- it will also be controlled exactly the same way as the electron is affected by the external applied voltage- however- it is opposite to the electrons motion-and much slower by its mass – however-its magnetic field will add to that of the electrons- this is standard knowledge of magnetic field generation of charged particles by way of electric fields-

Thus, in this thought experiment- within this state of dielectric breakdown- I like to say the lightning bolt- can you now see the oscillation of the charged particles by such a large magnitude of the applied voltage plates- can you now see the probability at a certain moment in time- particularly when the applied ac signal is greatest- that all magnetic fields of charged particles in transit can have the power to be magnetically pinched-

Magnetics have push and pull- it is well known that high frequency causes a constriction upon electrical currents- whether in copper lines and forcing such to the surface- or in plasma and made use to constrict it - high frequency is known to constrict the currents- thus- do you see the forces I am referring to as this dielectric breakdown has occurred upon the most perfect fuel of the heavens- water- this oxygen atom holds the hydrogen for us- we shall apply a high voltage(high is relative- I have stated 750kv in the vicinity of the hydrogen for fusion- I come to this by way of understanding the beta decay of a free neutron- if it can disintegrate- it can come together- made into a deuteron- then made into helium)-

Thus –we make helium and oxygen is then unbound and must also be recaptured- this is easy in expansion tanks- the byproducts are helium and oxygen-truly noble-

To increase the q of this system- a simple toroidal magnet as used in tokamaks for plasma control may be used for additional pinching-

I hope this helps- however- you must always think of the applied electric field and its magnetic inducement upon the charged particles- we will stably run the tesla coil upon the dielectric of water itself(more appropriately heavy water) and turn up the voltage for greater acceleration speeds and thus magnetic fields- we can control pressure in this vessel- we can control electrode spacing and so much more-

that is my thought experiment for you to understand it is my intention to use all input energy by way of oscillating a very large electric field which will induce the proper magnetic flux density in units of Tesla for a pinching- not established hot fusion whereby all energy input is chaotic and the probability of fusion comes by way of statistics from a gas equation- no- high voltage and high frequency in an orderly manner by the construction of a man made lightning bolt with controllable parameters inside an existing pressurized nuclear reactor-everything is off the shelf- the world will run on steam power again globally-from trains-factories-ships-and all power plants- i offer the Watt steam engine again- not with two sticks to make fire and boil the water- but electromagnetic fusion with two electrodes to induce fusion of hydrogen isotopes to boil water and make steam- it is absolutely perfect.

To the CEO's of nuclear power plants-or pressurized coal fired- you have spare reactors in the back yards of your plants- take a 1 million volt Tesla coil and do what I have said- you will measure its fusion reactions and you will say to yourself-hmmmm- I have found the road to safe clean energy.

One last note about the ideology exhibited towards Lightning Fusion process as found with a controlled charged particle acceleration such as cathode ray scopes and or the standard picture tube television.

What happened to the picturetube?

I was strolling thru walmart other day and I see no picturetubeTvs anymore-only lcds and plasmas- they dont make the glass picture tube anymore- the car has replaced the horse.
old school picturetube

Fascinating to see this. to see so much change in a generation. change is constant we all know for so much to be discovered.

I live several miles from the first transistor assembly plant on earth in Allentown pa. built in 1951- they closed the doors couple years ago and the property is for sale now.
The first transistor invented in new jersey and built in Allentown pa. for telephones and so much more-my father worked in the chemical dept etching chips for 30 years

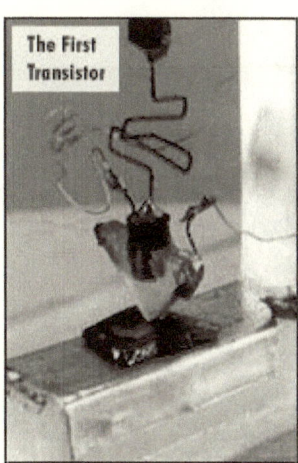

The First
Transistor

copyright: Lucent / Bell Labs

and ofcourse Bethlehem steel a town away with rusting plants from another time not so long ago with memories of the golden gate

and recently I hear about the big three and maybe their demise-as my 74 firebird formula still does donuts as it sits in my driveway - what is going on in America?- hmmm

back to the television- the picture tube-- this device was somewhat invented by William Crooks in the 1890's as scientists first learned to control with electricity how to shoot an electron thru a vacuum(the evacuated air from inside a picture tube) of primitive glass tubes in order for it to hit the other side and make it glow- well there aims were more than that - but for the predecessor of television-lets just say with the ability to make a little glow spot on a screen will then give you ability to make anything glow with control anywhere on a screen-and walluh- half a century later they figure out television- that was Philo Farnsworth.

a picture of William Crooks

but lets go back to William Crooks- and his tubes in which many called crooks tubes at the time- his work was fascinating for many eyes and minds- for Einsteins special theory of relativity as he tried to make sense of the mass increase as they accelerated the electron in the tubes- for Nikola Tesla as he took accelerating charged particles to a whole new level with his Tesla coils- a device with the ability to make man made lightning bolts-

a picture of Nikola Tesla

at this point, let me say even at the time of crooks in 1890's, many knew that to push an electron-or any charged particle- it will gain a magnetic field around itself as it moves- thus, with simple magnetics or electric coil around the path of flying electron-you can guide it-ever play with magnets -they have push and pull and so we can push and pull and guide this little electron as it goes from one side of the tube to another- but now with precision- and for the television later- being able to scan back and forth and up and down like a laser beam modulating the intensity of the electron beam- the glow on the screen varied to whatever you wanted- A PICTURE !

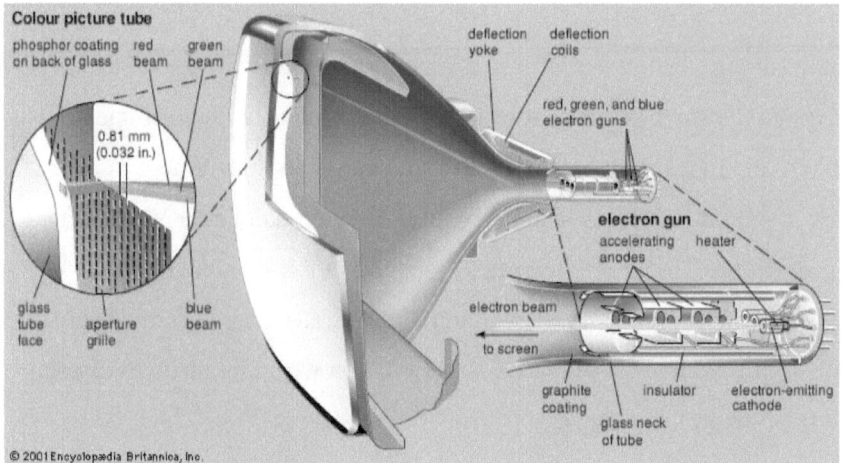

So this beautiful device called the picturetube television from ancient components of the birthplace of electricity and relativity is now being retired.

from pyramids to transistor plants-we move on with technology and life itself for something better we hope.

for me-I was born and raised in the Lehigh valley-it is the home to the American industrial revolution with its coal since 1700's , I've seen major industry pass right before my very own eyes and I am grateful and privileged to have experienced it.

I believe I have found a new way to make unlimited clean energy for the 21st century- a very simple process- just like the picture tube television- or cathode ray scope those in the science like to call it- with the ability and insight that we can create man made lightning bolts with nikola tesla coils- pushing and guiding charged particles just like the little electron in the tube- but this time inside water itself- very special type of water- to push the particles back and forth like the tide- like alternating current- but guiding and controlling it with external magnetic as we did with the little ole television picturetube- in the ideology built from Einsteins theory of special relativity as particles get accelerated this science come forth-

In this 21st century where many are aware of a strong need for clean energy- I hope and pray my deeds in science will end the global energy crisis and restart America and the world to greater heights-

I hope what has been given forth in this paper is enough for a reader to grasp and expand upon. For what is at stake is nuclear fusion with an abundant source of fuel to meet global demands and clean emissions. Energy is not a luxury; it should be a right of all citizens as should health care and quality living standards. Energy for warmth, light, and mobility make our lives far more comfortable and it is to that attainment we must pursue for not only ourselves but the lives of our children to live in a world free from pollutants and loss of natural resources.

I hope what has been given forth in this paper is enough for a reader to grasp and expand upon. For what is at stake is nuclear fusion with an abundant source of fuel to meet global demands and clean emissions. Energy is not a luxury; it should be a right of all citizens as should health care and quality living standards. Energy for warmth, light, and mobility make our lives far more comfortable and it is to that attainment we must pursue for not only ourselves but the lives of our children to live in a world free from pollutants and loss of natural resources. - May the light of God shine upon all our actions for the betterment of mankind.

Sincerely

Solomon Sami Azar

Noblefuse

December 25 2008